A Study In The Elements

By

Ian Beardsley

ISBN: 978-1-329-15741-5

Having worked with one of the most prominent Gypsy Shamans in the world today, and having had the good fortune to begin my tabla lessons and astronomy lessons with some of the greatest minds in those fields, we find that their wisdom lends itself to a form of archaeology that centers around the applied sciences.

Through the stories relating to certain individuals of vital character, we may in the end find that the naturally occurring elements were not just made for humans, but that their intended uses are written in their physical characteristics by an unknown hand.

Pi

The area of a circle is one half r time C, where r is its radius and C is its circumference. We immediately see that the ratio of the circumference to the diameter of a circle becomes impor-tant. It is the constant pi.

$$A = \frac{1}{2} rc$$
$$c = 2\pi r$$
$$A = \pi r^2$$

If we take a regular hexagon, which is a six-sided polygon with each side equal in length, and take each side equal to one, then if each side is one, so is the line drawn from each corner of the regular hexagon to its center, because it is made of equilateral triangles. And if we inscribe it in a circle, we can say the perimeter is close to the circumference of the circle and the line from each corner to the center (called a radius) is the same as the radius of the circle, then we have the ratio of the perimeter to the diameter is an approximation to pi and is 6/2 =3.00. If we increase the number of sides of the regular polygon, the perimeter comes closer and closer to the actual circumference of the circle and our value for pi becomes more accurate. If we increase the number of sides of the regular polygon enough times, we find that to three places after the decimal, pi is 3.141:

$$\pi = 3.141$$
$$A = \frac{1}{2} rc$$
$$D = 2r$$
$$c = 2\pi r$$
$$A = \pi r^2$$

The Golden Ratio

Let us draw a line and divide it such that the length of that line divided by the larger section is equal to the larger section divided by the smaller section. That ratio is The Golden Ratio, or phi:

$$\frac{a}{b} = \frac{b}{c}$$

$$a = b + c$$

$$c = a - b$$

$$a(a - b) = b^2$$

$$a^2 - ab = b^2$$

$$a^2 - ab - b^2 = 0$$

$$\left(\frac{a}{b}\right)^2 - \frac{a}{b} - 1 = 0$$

$$\left(\frac{a}{b}\right)^2 - \frac{a}{b} = 1$$

$$\left(\frac{a}{b}\right)^2 - \frac{a}{b} + \frac{1}{4} = \frac{5}{4}$$

$$\left(\frac{a}{b} - \frac{1}{2}\right)^2 = \frac{5}{4}$$

$$\frac{a}{b} = \frac{\sqrt{5} + 1}{2} = 1.618...$$

The arithmetic mean is the midpoint, c, between two extremes a, and c:

$$b = \frac{a + c}{2}$$

The harmonic mean is not necessarily the midpoint between two extremes but is the value that occurs most frequently:

$$b = \frac{2ac}{a + c}$$

The geometric mean, b, between a and c, is the side of a square that has the same area as a rectangle with sides a and c:

$$b = \sqrt{ac}$$

The following relationship holds:

$$a : \frac{a + c}{2} :: \frac{2ac}{a + c} : c$$

Data For The Planets

planet	Orbit (O)	Radius (R)	Mass (M)
mercury	0.387099	0.382	0.0558
venus	0.723332	0.949	0.8150
earth	1.000000	1.000	1.0000
mars	1.523691	0.532	0.1074
jupiter	5.202803	11.27	317.893
saturn	9.53884	9.44	95.147
uranus	19.1819	4.10	14.54
neptune	30.0578	3.88	17.23

O for Earth = 1.495979E13 cm   R for Earth = 6,378 km   M for Earth = 5.976E27 g

Earth-Moon Separation: 3.84E10 cm
Solar Radius: 6.9599E10 cm

Molar Mass of Gold: Au = 196.97
Molar Mass of Silver: Ag = 107.87

Saturn (minimum distance from sun) = 9.014 AU = 1.348E9 km
Jupiter (minimum distance from sun) = 4.951 AU = 7.409E8 km

Jupiter (maximum distance from the sun): 5.455 AU ~ 5.4 Astronomical Units

69

Amarjit

An Indian Tabla Set

I found myself in the medioambiente (atmosphere) of an Indian ethnomusicologist from the University of Delhi, taking tabla lessons. First he explained to me that he was not an idle man, that he had many students and taught from five in the morning to 6 in the evening, and that during that time he prepared a special soup for all those that were his students. One of the first things he told me was that in India there are many false gurus, that will not really teach you, but that this was not his consciousness, that he would really teach me. He said he would put me on a program of learning to play Bahjrans and Kirtans, rythms of 6 and 7 which fall under the category of Guzals, or Indian romantic music, but that he would be playing and composing temple music, called tin tal, which was the cycle of 16 considered the highest and most spiritual form of North Indian Classical music.

The training began with the history of the tabla which has its origins in the mridangam. It was the Muslim King in India, Amir Kusuro, who took the mridangam, which was closed on both ends, the left side played with the left hand and the right side played with the right hand, and broke it into two, the Dayan and Bayan, with the Dyan being the high tones and the Bayan being the low tones played with right and left hands respectively. In the center of each is a circle of dry ink that allows the drums to be tuned to precise pitches. The ink is rubbed into the tabla, as was explained to me, with a stone that floats on water and glows like a cat's eye and only exists in a few secret, undisclosed locations, only known to tabla makers. Amarjit, that was his name, had made it a point of telling me that among the rhythms I would be learning was a cycle of seven and one half and a cycle of 13 1/2. I find that interesting. If a person considers each beat of one half a beat of one, then that is a cycle of 15. It was the Gypsy Shaman, Manuel, who first pointed out to me that 15 was of primary importance, and as a scientist, I can't help but think in reference to that, the earth rotates through 15 degrees in an hour, and the most abundant element in the earth's atmosphere is nitrogen which is in chemical group 15 in the periodic table. Let us multiply Amarjit's 7 1/2 by the 16 of his tin tal. It is 120. 120 are the degrees in the angles of a regular hexagon, an equal angled, equal sided polygon with six sides. Let us subtract 120 from the 360 degrees that are in a circle and divide the result by that same 360 and then add the result to one:

$$360-120 = 240$$
$$240/360 = 2/3$$
$$2/3+1 = 5/3$$

This is the value that represents the yang of the cosmic yin and yang that came to us from the Gypsy Shaman, Manuel, that represents six-fold symmetry, or the physical aspects of nature, like snowflakes. The biological aspects are in five-fold symmetry, derived as above:

$$360/5 = 72$$
$$360-72 = 288$$
$$288/360 = 4/5$$
$$4/5 + 1 = 9/5 = 1.8$$

Let us divide Amarjit's stressed cycle 13.5 by 7.5. We find it is 1.8, which equals the yin of 9/5 that is representative of the organic aspects of nature to which the Gypsy Shaman, Manuel guided us in my story Gypsy Shamanism and the Universe, which I will present following the story we are telling now.

After my tabla lesson, I left the room and just as I came out, several people from India were coming into the house. I noticed in the living room was lots of clothing and art from India. I was introduced to these people, who obviously ran a store, and they told me they were just coming back from an interactive convention between Indians and Mexicans. The interchange was one between ideas in the cooking of Indian food and Mexican food. They were all wearing name tags that said on them, "Friendly Amigo".

Later I met with Amarjit and he took me to a music store to give me a lesson in buying instruments. On our way back, with his student driving, me in the front seat, Amarjit laid stretched out on the back back seat telling me that the store owner's refusal of our price offer for a crude guitar indicated that he was "A very greedy man and would not get far in life". At some point I told Amarjit that I had dreams of him giving me tabla lessons. He told me he could communicate with me in this way.

Upon learning that God told me the Gypsy Shaman, Manuel, always second guesses him, and Manuel telling me that because of this, he goes out into the world to do God's work for him at his request, Amarjit and his students were going to change their course from one of merging with God, to one of merging with Manuel.

Ian Beardsley
May 15, 2015

Manuel

Gypsy Shamanism And The Universe

I wrote a short story last night, called Gypsy Shamanism and the Universe about the AE-35 unit, which is the unit in the movie and book 2001: A Space Odyssey that HAL reports will fail and discontinue communication to Earth. I decided to read the passage dealing with the event in 2001 and HAL, the ship computer, reports it will fail in within 72 hours. Strange, because Venus is the source of 7.2 in my Neptune equation and represents failure, where Mars represents success.

Ian Beardsley
August 5, 2012

Chapter One

It must have been 1989 or 1990 when I took a leave of absence from The University Of Oregon, studying Spanish, Physics, and working at the state observatory in Oregon -- Pine Mountain Observatory—to pursue flamenco in Spain.

The Moors, who carved caves into the hills for residence when they were building the Alhambra Castle on the hill facing them, abandoned them before the Gypsies, or Roma, had arrived there in Granada Spain. The Gypsies were resourceful enough to stucco and tile the abandoned caves, and take them up for homes.

Living in one such cave owned by a gypsy shaman, was really not a down and out situation, as these homes had plumbing and gas cooking units that ran off bottles of propane. It was really comparable to living in a Native American adobe home in New Mexico.

Of course living in such a place came with responsibilities, and that included watering its gardens. The Shaman told me: "Water the flowers, and, when you are done, roll up the hose and put it in the cave, or it will get stolen". I had studied Castilian Spanish in college and as such a hose is "una manguera", but the Shaman called it "una goma" and goma translates as rubber. Roll up the hose and put it away when you are done with it: good advice!

So, I water the flowers, rollup the hose and put it away. The Shaman comes to the cave the next day and tells me I didn't roll up the hose and put it away, so it got stolen, and that I had to buy him a new one.

He comes by the cave a few days later, wakes me up asks me to accompany him out of The Sacromonte, to some place between there and the old Arabic city, Albaicin, to buy him a new hose.

It wasn't a far walk at all, the equivalent of a few city blocks from the caves. We get to the store, which was a counter facing the street, not one that you could enter. He says to the man behind the counter, give me 5 meters of hose. The man behind the counter pulled off five meters of hose from the spindle, and cut the hose to that length. He stated a value in pesetas, maybe 800, or so, (about eight dollars at the time) and the Shaman told me to give that amount to the man behind the counter, who was Spanish. I paid the man, and we left.

I carried the hose, and the Shaman walked along side me until we arrived at his cave where I was staying. We entered the cave stopped at the walk way between living room and kitchen, and he said: "follow me". We went through a tunnel that had about three chambers in the cave, and entered one on our right as we were heading in, and we stopped and before me was a collection of what I estimated to be fifteen rubber hoses sitting on ground. The Shaman told me to set the one I had just bought him on the floor with the others. I did, and we left the chamber, and he left the cave, and I retreated to a couch in the cave living room.

Chapter Two

Gypsies have a way of knowing things about a person, whether or not one discloses it to them in words, and The Shaman was aware that I not only worked in Astronomy, but that my work in astronomy involved knowing and doing electronics.

So, maybe a week or two after I had bought him a hose, he came to his cave where I was staying, and asked me if I would be able to install an antenna for television at an apartment where his nephew lived.

So this time I was not carrying a hose through The Sacromonte, but an antenna.

There were several of us on the patio, on a hill adjacent to the apartment of The Shaman's Nephew, installing an antenna for television reception.

Chapter Three

I am now in Southern California, at the house of my mother, it is late at night, she is a asleep, and I am about 24 years old and I decide to look out the window, east, across The Atlantic, to Spain. Immediately I see the Shaman, in his living room, where I had eaten a bowl of the Gypsy soup called Puchero, and I hear the word Antenna. I now realize when I installed the antenna, I had become one, and was receiving messages from the Shaman.

The Shaman's Children were flamenco guitarists, and I learned from them, to play the guitar. I am now playing flamenco, with instructions from the shaman to put the gypsy space program into my music. I realize I am not just any antenna, but the AE35 that malfunctioned aboard The Discovery just before it arrived at the planet Jupiter in Arthur C. Clarke's and Stanley Kubrick's "2001: A Space Odyssey". The Shaman tells me, telepathically, that this time the mission won't fail.

Chapter Four

I am watching Star Wars and see a spaceship, which is two oblong capsules flying connected in tandem. The Gypsy Shaman says to me telepathically: "Dios es una idea: son dos". I understand that to mean "God is an idea: there are two elements". So I go through life basing my life on the number two.

Chapter Five

Once one has tasted Spain, that person longs to return. I land in Madrid, Northern Spain, The Capitol. The Spaniards know my destination is Granada, Southern Spain, The Gypsy Neighborhood called The Sacromonte, the caves, and immediately recognize I am under the spell of a

Gypsy Shaman, and what is more that I am The AE35 Antenna for The Gypsy Space Program. Flamenco being flamenco, the Spaniards do not undo the spell, but reprogram the instructions for me, the AE35 Antenna, so that when I arrive back in the United States, my flamenco will now state their idea of a space program. It was of course, flamenco being flamenco, an attempt to out-do the Gypsy space program.

Chapter Six

I am back in the United States and I am at the house of my mother, it is night time again, she is asleep, and I look out the window east, across the Atlantic, to Spain, and this time I do not see the living room of the gypsy shaman, but the streets of Madrid at night, and all the people, and the word Jupiter comes to mind and I am about to say of course, Jupiter, and The Spanish interrupt and say "Yes, you are right it is the largest planet in the solar system, you are right to consider it, all else will flow from it."

I know ratios, in mathematics are the most interesting subject, like pi, the ratio of the circumference of a circle to its diameter, and the golden ratio, so I consider the ratio of the orbit of Saturn (the second largest planet in the solar system) to the orbit of Jupiter at their closest approaches to The Sun, and find it is nine-fifths (nine compared to five) which divided out is one point eight (1.8).

I then proceed to the next logical step: not ratios, but proportions. A ratio is this compared to that, but a proportion is this is to that as this is to that. So the question is: Saturn is to Jupiter as what is to what? Of course the answer is as Gold is to Silver. Gold is divine; silver is next down on the list. Of course one does not compare a dozen oranges to a half dozen apples, but a dozen of one to a dozen of the other, if one wants to extract any kind of meaning. But atoms of gold and silver are not measured in dozens, but in moles. So I compared a mole of gold to a mole of silver, and I said no way, it is nine-fifths, and Saturn is indeed to Jupiter as Gold is to Silver.

I said to myself: How far does this go? The Shaman's son once told me he was in love with the moon. So I compared the radius of the sun, the distance from its center to its surface to the lunar orbital radius, the distance from the center of the earth to the center of the moon. It was Nine compared to Five again!

Chapter Seven

I had found 9/5 was at the crux of the Universe, but for every yin there had to be a yang. Nine fifths was one and eight-tenths of the way around a circle. The one took you back to the beginning which left you with 8 tenths. Now go to eight tenths in the other direction, it is 72 degrees of the 360 degrees in a circle. That is the separation between petals on a five-petaled flower, a most popular arrangement. Indeed life is known to have five-fold symmetry, the physical, like snowflakes, six-fold. Do the algorithm of five-fold symmetry in reverse for six-fold symmetry, and you get the yang to the yin of nine-fifths is five-thirds.

Nine-fifths was in the elements gold to silver, Saturn to Jupiter, Sun to moon. Where was five-thirds? Salt of course. "The Salt Of The Earth" is that which is good, just read Shakespeare's "King Lear". Sodium is the metal component to table salt, Potassium is, aside from being an important fertilizer, the substitute for Sodium, as a metal component to make salt substitute. The molar mass of potassium to sodium is five to three, the yang to the yin of nine-fifths, which is gold to silver. But multiply yin with yang, that is nine-fifths with five-thirds, and you get 3, and the earth is the third planet from the sun.

I thought the crux of the universe must be the difference between nine-fifths and five-thirds. I subtracted the two and got two-fifteenths! Two compared to fifteen! I had bought the Shaman his fifteenth rubber hose, and after he made me into the AE35 Antenna one of his first transmissions to me was: "God Is An Idea: There Are Two Elements".

It is so obvious, the most abundant gas in the Earth Atmosphere is Nitrogen, chemical group 15 and the Earth rotates through 15 degrees in one hour.

The Bronze Age

Often the one thing you are looking for is the one thing that was left out of the story.

If you are an archaeologist you understand that gold and silver were important to early civilizations, especially to be used for ceremonial jewelries. But, you would also know that copper was used earlier and more as it is a soft and malleable metal that can be worked without being heated, pounded out into flat sheets.

Copper (Cu) used tin (Sn) as an alloying metal to make bronze, which was the beginning of the Bronze Age in Mesopotamia around 3500 BC.

These elements are the elements left out of Manuel's and Amarjit's stories, and so are just what are being suggested. Today the alloying metal for bronze is zinc (Zn). Let us look at the ratio of the molar masses of tin to zinc:

$$Sn/Zn = 118.71/65.39 = 1.8154 \sim 1.8 = 9/5$$

It is the nine-fifths around which our stories have been centered.

Ian Beardsley
May 15, 2015

Hand pounded copper ashtray demonstrating its malleability.

Two works in silver, one in gold, demonstrating its use for ceremonial and spiritual purposes.

Dr. James C. Kemp

When I was a physics student at the University of Oregon, I had the good fortune of be-ing taken in at the state observatory, Pine Mountain Observatory, and to work under its head, Professor James C. Kemp.

I have decided to write about him in this work, because of it being centered around peo-ple of vital character, and, if anyone was a man of dynamic character, it was Dr. Kemp.

First of all, to describe him, he was tall, thin, and handsome, and wore in the cold, snowy, sub-zero temperatures of Pine Mountain in the Winter, a parka, with wolverine fur around the hood, short pants, and went bare foot in the snow, with a corn cob pipe hanging out of his mouth. On one cold, Winter, starry night, standing between the tele-scope domes, looking at the starry sky, he told me: "We are not astronomers, we are physicists; the universe is a laboratory for energies higher than can exist on earth to be studied". That was the field in which he got his Phd, high energy physics, at The Uni-versity of Berkley. The amazing thing was that physics is not what he studied as an un-dergraduate—he studied Slavic Languages. How he managed to make the jump to graduate physics without first getting an undergraduate degree in the subject, is a mys-tery to me. But his study of Slavic Languages—He was fluent in Russian,-- began as early as high school, but not in America. As a child he and his mother moved to Mexico, and it was there that he went to the American School. His teacher there was Russian, and he told me she used to teach him Russian on their free time, that he had a great love for it. But it did not stop there. When as a young man, he came back to America, and he joined the Navy, and it was there that he came to have a Russian girl friend, through whom he became even more fluent in the language. After leaving the Navy, his education was paid for by United States Government. Though he did put him self through school repairing people's televisions, as well. He was studying Russian at Berkley and you might be wondering how he knew how to repair televisions. It goes back to his childhood in Mexico. He told me he loved doing electronics as a child in Mexico. He said he ran around the streets of Mexico, rummaging electronics components wherever he could find them so he could build a two-way radio for talking to North America. Which he did. He was then, as a child, a self-taught electronics engineer. Perhaps that had something to do with him being able to go out of Russian studies at Berkley, and directly into graduate level physics.

I always felt he much more enjoyed being at the observatory, which was some 200 miles east of the university in the high desert of Oregon, with its sage, ponderosa pine, herds of antelope, and pristine skies. When one left the university, one took a windy highway with no traffic, along a raging McKenzie River, that flowed through dense enchanted forests. Somewhere along that highway, there was one secluded restaurant called, Mom's Homemade Pie. When I didn't make the commute myself, in my 1976 Datsun pickup, and went with him, we would always stop there and he we buy me a cheeseburger and we would top it of with some of Mom's Homemade Pie.

The second we got to the observatory, he would take off his sandals, so he could go barefoot, whether there was snow or shine. He was told me he climbed the South Sister, barefoot. The south sister is one of the Mountains of the Cascade Mountain Range that you must drive over when leaving the University and heading to the observatory. He did not want to ever wear shoes at all— I think it was part of his free, childlike spirit— but had to when he was at his office at the department of physics. Well he didn't exactly have to wear shoes, sandals would suffice and that was exactly what he did. However, there was only one kind of sandal he would wear, and that was leather sandals from Mexico, called Huarachis, and he always brought a few pairs back stateside when he went to Mexico on astronomy business. Huarachis are braided leather, but he told me the soles were made from car tires, and that you can get 200,000 miles on them.

What about music? Well on cloudy, snow nights, when we could not work with the telescopes, we would sit up all night in the warm residence, waiting for a break in the cloud cover, and he would play for me a record he liked a lot, it was by Joni Mitchell. It was not the only record we listened to, but he would play for me a record he bought at the airport in the Soviet Union. He worked with Russian astronomers, and when he was in Russia, I would stay behind and run the telescopes stateside. So, his earlier studies in Russian served him well in the sciences, as the Soviet Union is one of the leaders in astronomical sciences. He would explain to me the lyrics on the record, which was called Moscow Nights.

But when he was with the telescopes, at night, in the cold snow, he listened to only one composer that I can think of, and that was the Finnish composer Sibelius. In retrospect, I see it as a wise choice. Personally, when I hear his work, it is the only time I don't feel I would rather be listening to Mozart, Bach, or Beethoven. I think he is perhaps not trumped by these geniuses because of his is bold, brave, and mysterious sound, like the ice lands of Finland that he depicted. His depictions of Finland perfectly described the snowy, Pine Mountain landscape where we worked.

Just what kind of work did we do at the observatory? While we studied the magnitude of light from a star, and its changes over time, what is known as photometry, the most important work was in something for which Professor Kemp built the measuring device, that was the most sensitive in the world at the time, because of his patented crystal, called polarimetry. Which is not just the study of how bright something is, but how its light is polarized, both in the directions of its vibration, and its percentage (how much is polarized, and in what direction). This enabled one to figure out the geometry of celestial systems. The amount of polarization is related to the reflective surface area of the object studied, and the direction of polarization, to the orientation of the object.

Professor Kemp once told me a secret: "You could in theory detect planets around a star with polarimetry, because they shine by reflected light and reflected light is polarized. Not only can you detect them, but you can determine their size and orbital orientation. Our telescopes are too small to do it, but my polarimeter would be sensitive enough to do it with a 200 inch telescope. With that, I can get down to a part in one million; that is just what is needed, observe:

"For a Jupiter sized planet (radius = 71880000 meters) at earth orbit (radius = 1.5E11 meters) we have

R squared/(r squared) = (71880000^2)/(1.5E11)^2=5E15/2.25E22 =

0.000,000,222"

The Iron Age

Following the Bronze Age came the Iron Age, around 1200 BC in Anatolia, where weapons and tools were made from iron. This age has continued until today and were are still in it. Anatolia is the Westernmost protrusion of Asia and it makes up the majority of the Republic of Turkey. It is bound to the North by the Black Sea, to the South by the Mediterranean Sea, and the Aegean Sea to the West. It is interesting to note that the ratio of the molar masses of Gold (Au) to Iron (Fe) is 3.5 because the luminosity of a star is exponentially related to its mass and that exponent is 3.5. The luminosity of a star of 100 solar luminosities then has a mass of 3.7 solar masses.

$\log(100) = 3.5 \log (M)$
$2/3.5 = \log (M)$
$0.57 = \log (M)$
$M = 3.7$

Steel is an alloy of iron and carbon and has a high tensile strength. In Indian music the tabla player is at odds with ascension because he or she has to put their hands to an animal skin drum head that was made by slaughtering an animal, which is a hinderance in hinduism to ascension through reincarnation. However this problem is not intrinsic to the sitar player, because the strings are made of steel which at once produces a tone conducive to spiritual ascension. The ratio of iron (Fe) to carbon (C) in molar masses is:

$Fe/C = 55.85/12.01 = 4.64696$

More About Nine-Fifths: Pi and Phi

The molar mass of gold to silver is nine to five, just like the ratio of the solar radius to the lunar orbital radius. The sun is gold in color and the moon is silver in color.

Pi and Phi

But is not nine-fifths a more dynamic number than what I have pointed out so far? Consider the golden ratio (denoted Φ called phi). And consider the ratio of the circumference of a circle to its diameter (denoted π called pi).

$$\Phi + \pi = 1.618 + 3.141 = 4.759$$

The four takes you around a circle 4 times, that is back to where you started, the fraction after 4, the 0.759, is the important part, it is what is left. Notice the seven is the average of nine and five and the 5 is the 5 of nine-fifths, and the nine is the nine of nine-fifths. We see that nine-fifths unifies the sum of pi and phi.

The earliest use of glass was the use of naturally occurring glasses like obsidian by stone age civilizations to make sharp tools. The first manufactured glass was made in coastal Syria, Mesopotamia or Egypt as early as 3000 BC, the earliest products being beads.

Glass is made by heating sand in oxygen, which means it is basically silicon dioxide.

$Si = 28.09$
$O = 16.00$

$SiO_2 = 28.09 + 2(16.00) = 28.09 + 32.00 = 60.09$

$Si/O_2 = 28.09/32.00 = 0.8778125$
$O_2/Si = 32.00/28.09 = 1.139$
$Si/O = 28.09/16.00 = 1.755625$
$O/Si = 16.00/28.09 = 0.569597722$
$Si/SiO_2 = 28.09/60.09 = 0.467$
$O/SiO_2 = 16.00/60.09 = 0.266267266$
$O/SiO_2 \sim 16/60 = 4/15$
$O_2/SiO_2 \sim 32/60 = 8/15$

Silicon, Phosphorus, And Boron

Back in 2005, as I did my research, I developed a different convention for rounding numbers than we use. I felt I only wanted to use the first two digits after the decimal in processing data using molar masses of the elements. This I did, unless a fourth digit less than five followed the third digit in my calculations, then, I would use the first three digits for greater accuracy. Now I am taking the introductory class in computer science at Harvard, online, CS50x. Working in binary, where all numbers are base two, I see that it was no wonder I got the results I did, on the first try when I wondered if the golden ratio conjugate, 0.618 to three places after the decimal would be in artificial intelligence (AI) since it is recurrent throughout life.

I was taking polarimetric data on the eclipsing binary Epsilon Aurigae at Pine Mountain Observatory in the 1980's, for which there was a paper in the Astrophysical Journal upon which my name appears as coauthor, while studying physics at The University of Oregon. As well I was studying Spanish, and in an independent study project through the Spanish Department, I left the University to live of among the caves of the Gypsies of Granada, Spain. In doing as such, I disappeared from the entire world, only to return from another kind of life finding the world was now a much different place. Around 2005, I enrolled in chemistry at Citrus College in Southern California, when I did the following:

If the golden ratio conjugate is to be found in Artificial Intelligence, it should be in silicon, phosphorus, and boron, since doping silicon with phosphorus and boron makes transistors.

We take the geometric mean between phosphorus (P) and Boron (B), then divide by silicon (Si), then take the harmonic mean between phosphorus and boron divided by silicon:

$$\sqrt{PB}\,/Si = \sqrt{(30.97)(10.81}\,/28.09 = 0.65$$

$$\frac{2PB}{P+B}\,/Si = \frac{2(30.97)(10.81)}{30.97+10.81}\,/28.09 = 0.57$$

Arithmetic mean of these two numbers: (0.65 + 0.57)/2 = 0.61
0.61 is the first two digits of the golden ratio conjugate.

Now the golden ratio conjugate is in the ratio of a persons height and the length from foot too navel, and is in all of ratios between joints in the fingers, not to mention that it serves in closest packing in the arrangement of leaves around a stem to provide maximum exposure to sun and water for the plant. Here we see that the golden ratio is not in artificial intelligence which is 0.62 to two places after the decimal, but that the numbers in its value are in artificial intelligence 0.61, which is 0.618 to three places after the decimal. That is, if we consider the first two digits in the ratio. If we consider the golden ratio conjugate to one place after the decimal, which is 0.6, then we say artificial intelligence does have the golden ratio in its transistors. I like to think of I, Robot by Isaac Asimov, where in one of that collection of his short stories, robots are not content with what they are, and need more: an explanation of their origins. They can't believe that they are from humans, since they insist humans are inferior. Or, I like to think of the ship computer HAL in 2001, he mimics intelligence, but we don't know if he is really alive. Perhaps that is why to two place after the decimal, AI carries the digits, but is not the value.

In any case, I have written a program called Discover that would enable one to process arithmetic, harmonic, and geometric means for elements or whatever, because someone, including myself, might want to see if there are any more nuances hidden out there in nature. I have already found something that seems to indicate extraterrestrials left their thumbprint in our physics. I even find indication for the origin of a message that would seem they embedded in our physics. That origin comes out to be the same place as the source of the SETI Wow! Signal, Sagittarius. The Wow! Signal was found in the Search For Extraterrestrial Intelligence and a possible transmission from ETs. But that is another subject that is treated in my book: All That Can Be Said.

The Elements According To The Ancient Greeks

An element is that which cannot be broken down further in a chemical process, such as heating. Compounds of elements can be broken down by heating or mixing with a liquid, but elements cannot. This is the definition of an element in chemistry. But to the Ancient Greeks, elements were classified as earth, air, water, and fire. But let us look at what these natural phenomena actually are in terms of chemistry and not its father, alchemy.

Earth is a solid, which is a phase of a substance wherein its atoms are not very movable because they are bound to one another due to it not being warm enough. Air is a phase of a substance for which it is a gas, which means its atoms are warm enough that they become excited enough that they are separated from one another. Water is a phase of H20 where it is a liquid, which means it is too cool for the its atoms to be excited enough to separate from one another, but too warm for its atoms to be so bound that they cannot hardly move. This means a liquid can take the shape of its container. Fire is the emission of light by a gas that has become excited by a reaction such as methane being burnt in oxygen. It is a self sustained reaction provided by its own heat. Aristotle proposed a fifth element called ether, that filled space. That makes five elements all together. Because their are five pythagorean solids, solids whose faces are the same regular polygons, which are equal sided, equal angled shapes like the equilateral triangle, the square, and the regular pentagon all with sides of equal length. The tetrahedron represented fire, the icosahedron represented water, the dodecahedron represented ether, the octahedron represented air, and the cube represented earth.

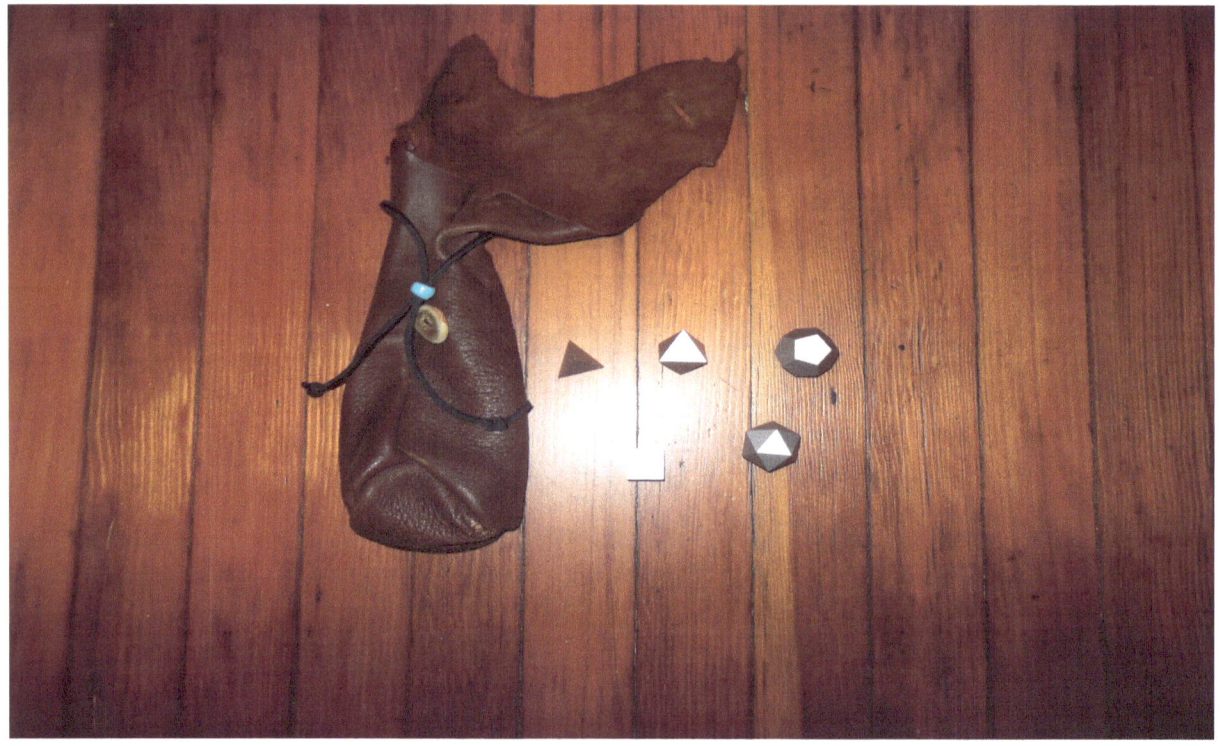

Top left tetrahedron, middle top octahedron, top right dodecahedron, bottom left cube, bottom right icosahedron.

Archaeology is in part the study of artifacts to understand civilizations and their cultures, even in contemporary times. These contemporary items could be considered artifacts.

This Medicine Pouch made by the author's brother, is reminiscent of many ancient cultures and peoples, not just American Indians, but Druids, and perhaps something an Ice Age hunter might have carried. Shown next to it is a Zuni Fetish.

The Author

www.ingramcontent.com/pod-product-compliance
Lightning Source LLC
Chambersburg PA
CBHW051111180526
45172CB00002B/869